Spotlight on Climate Change

Causes of Climate Change

Tracy Sue Walker

Lerner Publications ◆ Minneapolis

For Mom and Dad

Lerner Publications Company
An imprint of Lerner Publishing Group, Inc.
241 First Avenue North
Minneapolis, MN 55401 USA

For reading levels and more information, look up this title at www.lernerbooks.com.

Main body text set in Adrianna Regular.
Typeface provided by Chank.

Editor: Cole Nelson

Library of Congress Cataloging-in-Publication Data

Names: Walker, Tracy Sue, author.
Title: Causes of climate change / Tracy Sue Walker.
Description: Minneapolis : Lerner Publications, [2023] | Series: Searchlight books - spotlight on climate change | Includes bibliographical references and index. | Audience: Ages 8–11 | Audience: Grades 4–6 | Summary: "Climate change affects everyone around the globe, but what causes it? Readers explore the science behind the greenhouse effect, where greenhouse gases come from, and what they can do to lower their own carbon footprints"— Provided by publisher.
Identifiers: LCCN 2021051344 (print) | LCCN 2021051345 (ebook) | ISBN 9781728457918 (library binding) | ISBN 9781728463902 (paperback) | ISBN 9781728461885 (ebook)
Subjects: LCSH: Climatic changes—Juvenile literature.
Classification: LCC QC903.15 .W368 2023 (print) | LCC QC903.15 (ebook) | DDC 363.738/74—dc23/eng/20211204

LC record available at https://lccn.loc.gov/2021051344
LC ebook record available at https://lccn.loc.gov/2021051345

Manufactured in the United States of America
1-50819-50158-2/15/2022

Table of Contents

Chapter 1

IT'S GETTING HOT IN HERE!

We see climate change in the news every day, from flooding and wildfires to protests in the streets. It affects many people through natural disasters, bad food-growing seasons, and heat waves. But what exactly is climate change and what causes it?

Young people around the world are protesting the actions causing climate change, such as burning fossil fuels.

Climate change is a complex problem caused by Earth's atmosphere rapidly warming. It is hard to see directly, but we can see the effects of climate change in the increasing number of intense storms and weather events each year. Understanding climate change and its causes is an important step in figuring out how to fix it.

Human Activity

Earth's temperature, wind, and rainfall fluctuate and change over time. But over 90 percent of scientists agree that these changes in climate are happening quicker than ever in human history. Scientists have proved that this rapid change is caused by human actions.

While many human activities contribute to climate change, the use of fossil fuels for energy is the greatest cause. When fossil fuels burn, they release greenhouse gases such as carbon dioxide. Humans release billions of tons of greenhouse gases into the atmosphere every year. The buildup of these gases results in climate change.

FLARE STACKS AT PETROLEUM REFINERIES RELEASE GREENHOUSE GASES INTO THE AIR.

The Greenhouse Effect

When you grow plants inside a greenhouse, the glass walls allow in heat from the sun. That heat is then trapped by the glass. Earth's atmosphere works in a similar way. When energy from the sun reaches Earth, the planet's atmosphere keeps most of it. Only 30 percent of the sun's energy escapes back into space. Greenhouse gases absorb and trap heat in Earth's atmosphere. This is called the greenhouse effect.

Greenhouse gases trap heat from the sun, keeping Earth warm.

STEM Spotlight

Climate change and global warming go hand in hand but are not the same. Climate change refers to changes in Earth's average temperature, rainfall, and weather patterns over a long period. Global warming is the long-term heating of Earth's climate. Climate change can be caused by both natural events and human activities. But current global warming is caused only by human activity.

The greenhouse gas carbon dioxide enters Earth's atmosphere when fossil fuels like coal and oil burn. Methane, fluorinated gases, and nitrous oxide are also greenhouse gases.

Greenhouse gases are not all bad. Without them, the world would be much colder. But high levels of greenhouse gases in the atmosphere can do great damage. Some greenhouse gases hold onto more energy than others, such as methane, and these gases contribute more to Earth's warming and climate change.

UNDERSTANDING FOSSIL FUELS

For over one hundred years, people have used and depended on fossil fuels. Burning fossil fuels produces energy. But it also releases greenhouse gases into Earth's atmosphere. This is the biggest cause of climate change in the world today.

What Are Fossil Fuels?

Fossil fuels are used for creating electricity, heating, and transportation. Millions of years ago, many prehistoric plants and animals became buried under layers of rock after they died. Over time, the plants and animals became fossils. Fossil fuels are the remains of these plants and animals. Examples are coal, oil, and natural gas.

Oil is found by drilling into the earth. At a rig site, a driller uses drilling equipment to make a hole deep in the earth. Other workers insert a pipe into the hole and pump oil through the pipe to the surface. Natural gas is also found by drilling into Earth's surface.

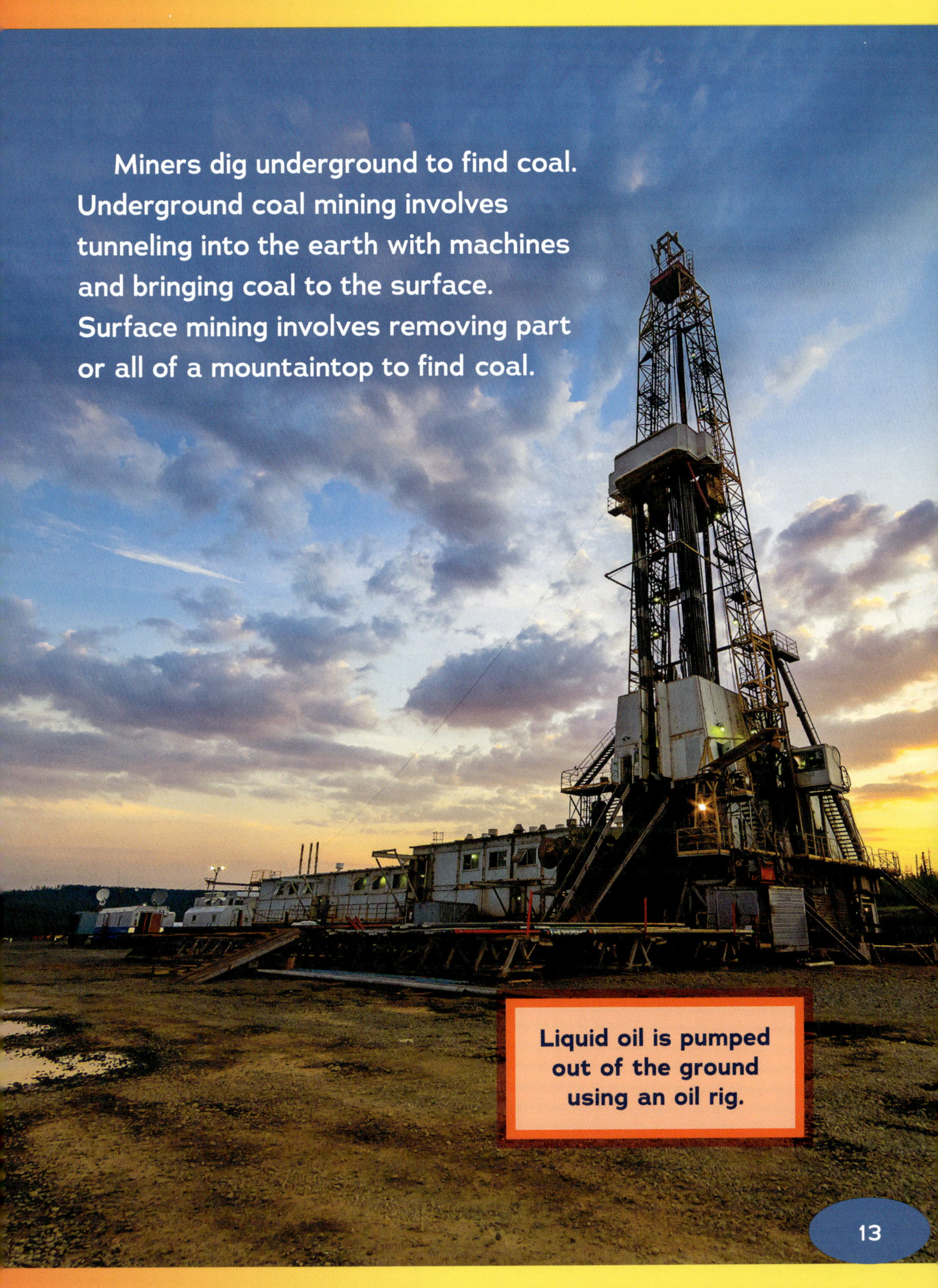

Miners dig underground to find coal. Underground coal mining involves tunneling into the earth with machines and bringing coal to the surface. Surface mining involves removing part or all of a mountaintop to find coal.

Liquid oil is pumped out of the ground using an oil rig.

Hydraulic fracturing, or fracking, is another way to mine for oil and gas. In fracking, drillers use water at high pressure to crack rocks underground and release fossil fuels.

Fossil fuels are an important part of many economies. Many countries still depend on selling fossil fuels and the jobs the fossil fuel industry provides. But renewable energy like solar power and wind power is getting cheaper to make, and more and more countries are switching away from fossil fuels.

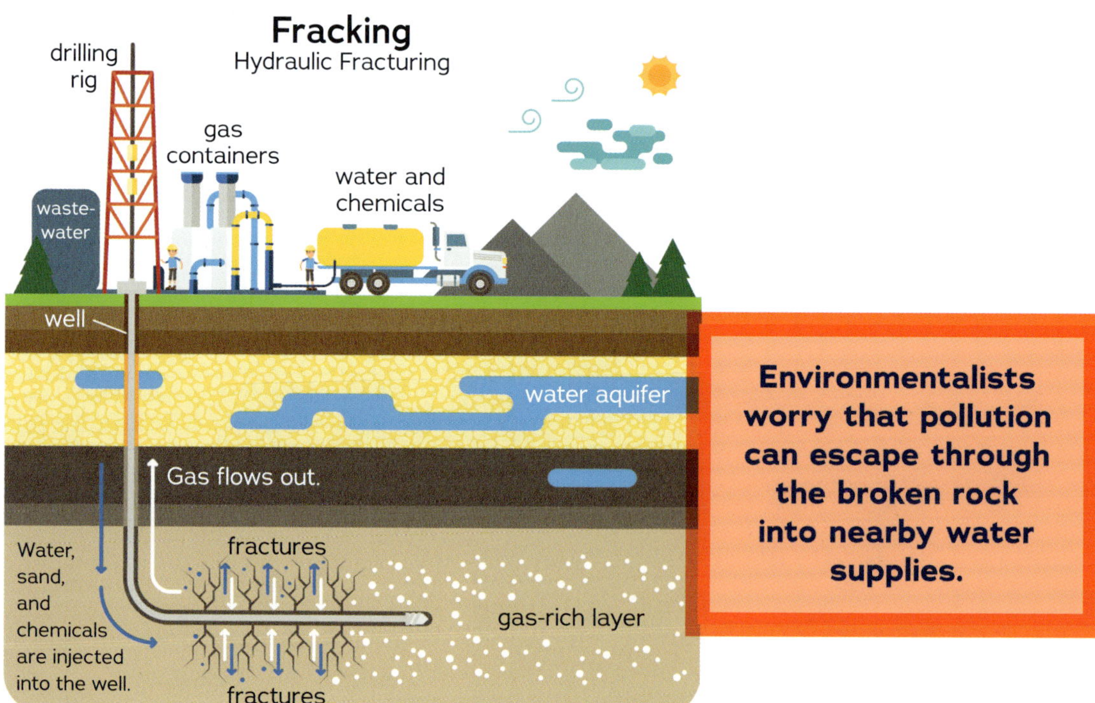

Environmentalists worry that pollution can escape through the broken rock into nearby water supplies.

Bill McKibben

Bill McKibben is an author and climate change activist. He is also a founder of 350.org, a grassroots climate movement that has groups in 188 countries and on every continent. The goal of 350.org is to get countries and businesses to stop using fossil fuels and replace them with renewable energy sources like sunshine and wind. McKibben was awarded the John Muir Award, the Sierra Club's highest honor, in 2011 for his work in mobilizing volunteers to fight climate change.

Bill McKibben in 2013

What Do Fossil Fuels Do?

Fossil fuels are used for many important jobs. In the past, many people burned coal to heat homes and move vehicles. Modern power plants burn coal to create electricity.

Oil, or petroleum, is used for transportation, heating homes, and electricity. Oil is an important ingredient in many of the plastics and chemicals we use daily. In the US, many power plants burn natural gas to create electricity.

CONTRIBUTING FACTORS

The burning and use of fossil fuels as energy sources is the primary cause of modern climate change. It's not the only cause. Many other human activities and behaviors contribute.

Deforestation

Deforestation is the clearing or burning of trees and forests. The harmful practice of clearing land for farms and houses is a key cause of climate change. Trees store a lot of carbon. The older the tree, the more carbon it stores. Clearing trees releases carbon dioxide into the atmosphere.

Since 1990, the number of trees cleared by humans equals an area about the size of South Africa. Protecting trees—especially the oldest and largest—is important to slowing climate change.

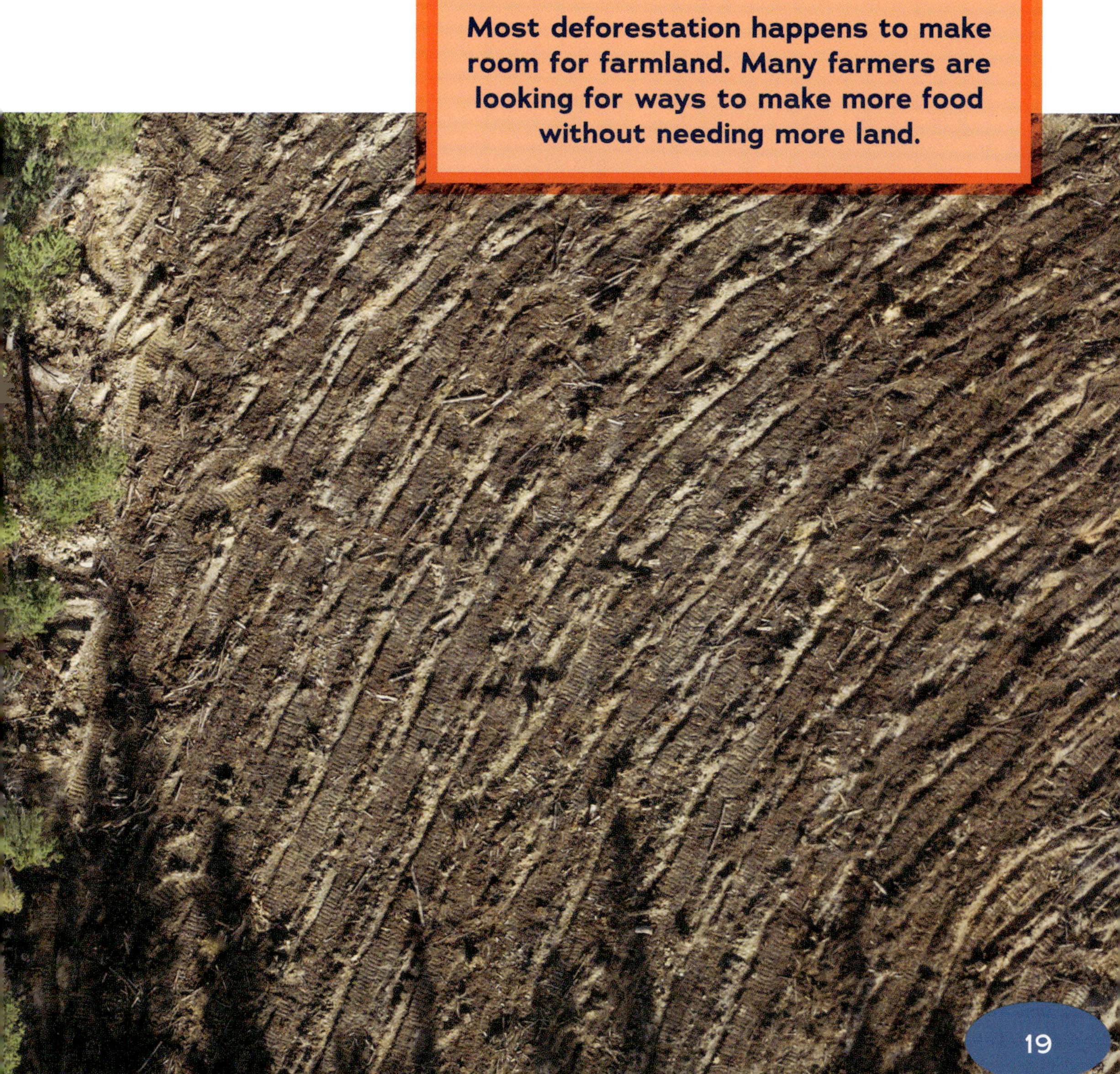

Agriculture and Livestock

Livestock has a huge impact on climate change. About 15 percent of all greenhouse gas emissions globally are caused by cows and sheep, largely from burping methane gas. Farmers around the world are also cutting down many trees to make room for cattle. Experts say swapping beef for a plant-based diet would help curb climate change.

What We Buy

Most clothes are manufactured with fibers made from fossil fuels. Clothes cost less than in the past, so people buy more. But clothes don't last as long as they used to, and more of them quickly end up in landfills. Billions of single-use plastics, like water bottles and bags, are used globally. They are produced using large amounts of oil. When we need to get rid of plastic products, we often have to burn them. This burning releases dangerous toxic chemicals into the air and water nearby.

Waste plastics and cloth products that don't easily degrade and cannot be recycled often end up in the ocean, which threatens sea life.

Transportation

Most cars, trucks, airplanes, and trains rely on burning fossil fuels like gasoline and diesel. This releases carbon dioxide into Earth's atmosphere. Transportation makes up 29 percent of the total greenhouse gas emissions in the US, the largest source of greenhouse gas emissions in the country.

Because airplanes release a lot of greenhouse gases, people around the world are committing to fly less.

STEM Spotlight

Electric vehicles (EVs) are powered by electricity. Instead of gasoline or diesel fuel, they use batteries that can be recharged at an electrical outlet. Conventional vehicles, like gasoline-powered cars and trucks, have high greenhouse gas emissions. EVs have much lower emissions.

EV popularity is growing. In 2020, about 1.8 million EVs were on the roads in the US. That's more than three times as many as there were in 2016.

Chapter 4

ADDRESSING CLIMATE CHANGE

Countries and communities can help fix climate change in many ways. The most important way is to use energy sources that protect Earth's atmosphere and reduce greenhouse gas emissions. We can help by contacting community leaders to make sure our cities are using renewable energy.

Reducing Our Carbon Footprint

Carbon footprints are the amount of greenhouse gases produced by humans. Many things contribute to your carbon footprint, like the food you eat, the clothes you buy, and the types of transportation you use. You can reduce your carbon footprint by turning off unused lights, eating less meat and more locally grown food, carpooling with other people or taking the bus, and recycling.

NOT WASTING FOOD AND EATING FOOD GROWN NEAR YOU CAN HELP LOWER YOUR CARBON FOOTPRINT.

Using Renewable Energy Sources

Renewable energy sources are created and replaced by nature. These energy sources include the sun (solar power) and wind. Unlike nonrenewable sources of energy such as fossil fuels, renewable energy is less harmful for Earth's environment. Renewable sources do not produce greenhouse gas emissions. Carbon footprints can be greatly reduced by using renewable energy.

Google and many other companies are investing in green energy like wind power for their energy needs.

Patricia Espinosa

Patricia Espinosa (*below*) is the executive secretary of the United Nations Framework Convention on Climate Change. Its main goal is to stop human activity from causing dangerous changes to Earth's climate.

Espinosa was born in Mexico in 1958. A diplomat since 1981, she has degrees in international law and international relations and specializes in climate change, sustainable development, and human rights.

Act Responsibly

When shopping, look for an Energy Star symbol. When a product receives an Energy Star symbol, it meets energy standards in the US. These products help reduce greenhouse emissions and climate change. Find Energy Star products at https://www.energystar.gov.

Purchasing products that are energy efficient, reducing your carbon footprint, and switching to renewable energy can help slow climate change. Responsible choices like these make a big difference and show you are doing your part to help protect the planet.

You Can Help!

There are many ways you can reduce your carbon footprint and help in the fight against climate change. Here are some ways to start:

At Home

- Turn off lights and unplug appliances when you're not using them.
- Switch to LED light bulbs since they use less energy and last longer than other bulbs.

What You Eat

- Eat plant-based meals at least one or two days every week.
- When you can, eat foods that have been produced locally in your community.
- Use washable dishes instead of disposable dishes and silverware.

On the Go

- If you usually travel by car, try walking, biking, or taking a train or bus instead.
- Carpool to help reduce the number of vehicles burning fossil fuels.

What You Wear

- Donate old clothes rather than throwing them away.
- Buy clothes you will wear many times, not just once or twice.

Glossary

carbon dioxide: a greenhouse gas made of carbon and oxygen

climate change: a change in the usual weather conditions of a place

deforestation: clearing land for other uses by cutting or burning trees or forests

electric vehicle: a vehicle that is powered by electricity

fossil fuel: a fuel containing carbon that is formed from prehistoric animal and plant remains

global warming: the increase in average temperature in Earth's atmosphere over a long period

greenhouse gas: a gas that traps heat in Earth's atmosphere

livestock: animals raised or kept for food, milk, or other products

Learn More

Kids against Climate Change
    https://kidsagainstclimatechange.co

Klein, Naomi. *How to Change Everything: The Young Human's Guide to Protecting the Planet and Each Other*. New York: Atheneum Books for Young Readers, 2021.

Kurtz, Kevin. *Climate Change and Rising Temperatures*. Minneapolis: Lerner Publications, 2019.

NASA Climate Kids
    https://climatekids.nasa.gov

National Geographic Kids: Climate Change
    https://kids.nationalgeographic.com/science/article/climate-change

Sharif-Draper, Maryam. *Climate Change*. New York: DK, 2020.

Index

Photo Acknowledgments

Image credits: DisobeyArt/Shutterstock.com, p. 5; Arthur Villator/Shutterstock.com, p. 6; Rex Wholster/Shutterstock.com, p. 7; VectorMine/Shutterstock.com, p. 8; NASA, p. 10; juninatt/Shutterstock.com, p. 12; Evgeny_V/Shutterstock.com, p. 13; VectorMine/Shutterstock.com, p. 14; AP Photo/Toby Talbot, p. 15; Jia Ji Yin/EyeEm/Getty Images, p. 16; Charlie Rogers/Getty Images, p. 18; GroanGarbu/Shutterstock.com, p. 20; neenawat khenyothaa/Shutterstock.com, p. 21; Greg Bajor/Getty Images, p. 22; Monkey Business Images/Shutterstock.com, p. 25; Steve Proehl/Getty Images, p. 26; Emily Macinnes/Bloomberg/Getty Images, p. 27; Muriel de Seze/Getty Images, p. 28.
Cover image: georgeclerk/Getty Images.